Anonymous

Life Beneath the Waves and a Description of the Brighton Aquarium

With numerous illustrations, including ground plan of the aquarium and view of the largest tank

Anonymous

Life Beneath the Waves and a Description of the Brighton Aquarium
With numerous illustrations, including ground plan of the aquarium and view of the largest tank

ISBN/EAN: 9783337095543

Printed in Europe, USA, Canada, Australia, Japan

Cover: Foto ©berggeist007 / pixelio.de

More available books at **www.hansebooks.com**

LIFE BENEATH THE WAVES

AND A

DESCRIPTION OF THE BRIGHTON AQUARIUM.

With Numerous Illustrations,

INCLUDING GROUND PLAN OF THE AQUARIUM

AND

VIEW OF THE LARGEST TANK.

LONDON:
TINSLEY BROTHERS, 18, CATHERINE STREET, STRAND.
1871.

PREFACE.

When I commenced this little work I did not intend to have it published until somewhat later in the year, when the Brighton Aquarium will be completed and open to the public. But after taking into consideration the fact that most people visit the seaside during the summer and autumn months, it does not seem to me advisable to delay the production of my endeavours to excite feelings of interest respecting the inhabitants of the ocean.

And I trust that my brief accounts of some of these submarine animals, and the description of the Brighton Aquarium in its finished state, will prove to be a combination sufficiently interesting to induce my readers to visit the above-named building (when it shall be ready for public inspection), not only from the desire of beholding its manifold architectural perfections and attractions, but also from the wish to become better acquainted with that portion of creation which constitutes

"LIFE BENEATH THE WAVES."

BRIGHTON, *July*, 1871.

LIST OF ILLUSTRATIONS.

Ground Plan of Aquarium	to face page	7
View of Largest Tank	,,	14
Dahlia Anemone	,,	24
Smooth Anemones	,,	26
Sabellæ	,,	30
Sea Worm—Serpulæ	,,	36
Scarlet Worm—Swimming Crab Changing its Shell	,,	40
Hermit Crab—Angled Crab	,,	46
Spider Crab—Nut Crab	,,	46
Spider Crab	,,	50
Helmet Crab—Swimming Crab	,,	54
Large Pecten	,,	58
Pecten—Dog Winkle—Pelican's Foot	,,	62
Trochus—Naticas	,,	64
Chitons—Venus	,,	66
Sea-mouse—Cowry and Pecten	,,	70
Dead Men's Fingers	,,	74
Webbed Starfish—Brittle Star	,,	78
Sea Urchin—Five-fingered Star	,,	82
Shrimp and Prawn	,,	86
Squat Lobster and Sea Slug	,,	90

NEW ROAD FACING SEA

WALL OF MARINE PARADE

REFERENCES.

A. ENTRANCE STEP DESCENDING FROM WEST END
B. ENTRANCE COURT
C. ENTRANCE HALL
D. DINING HALL
E E E. GROINED ARCH CORRIDORS FOR SPECIMENS IN TANKS ON EACH SIDE
F F F TANKS FOR SPECIMENS
C C CONSERVATORY
H H SPACE FOR ROCKWORK
I I SPACE FOR SMALL TANKS
J J ENGINE ROOM STORES &c

— Scale of feet —
10 20 40 60 100   200   300 feet.

LIFE BENEATH THE WAVES.

PART I.

The indifference which exists amongst people in general respecting the inhabitants of the sea, is without doubt a fact to be greatly regretted; but at the same time, surprise can scarcely be mingled with regret, when we take into consideration the few opportunities which have been offered to the public in this country, for acquiring any information respecting Life Beneath the Waves. For whilst children at school, and at home, are taught almost from their infancy to feel interested in the observance, and study of animals, birds, insects, flowers, and trees, no effort, generally speaking, is made to excite similar desires for information with regard to the denizens of our oceans and rivers.

And again, whilst facilities abound both in public and private for rendering ourselves familiar with, at any rate, the *outward* forms of creatures which live on the earth, or in the air, from the majestic Lion king, and fierce sovereign of the feathered tribes, to the insignificant field-mouse, and crawling reptile, very few opportunities have existed, up to the present time, for the acquisition of a corresponding amount of information respecting the living treasures of the deep.

The result of this state of things cannot, to say the least, be deemed satisfactory; for although in some cases "ignorance" *may* be "bliss," the exceptions to this rule are far more numerous than the corroborations thereof.

The plea of lack of opportunity for acquiring information respecting the inhabitants of the sea, can, however, no longer be as justly brought forward as hitherto, for the construction of a magnificent public Aquarium in the fashionable town of Brighton, effectually fills the gap so long left vacant, and supplies the opportunity so long left wanting, both for scientific and unscientific persons to render themselves acquainted, not only with the outward forms of the inhabitants of the ocean, but also for studying at leisure the various " manners and customs " of the same. For the object of the Brighton Aquarium is not merely to please the eye, and enable fashionable visitors to while away time in a novel and agreeable manner; it is intended to *instruct* as well as to delight, and at once to excite and gratify a thirst for information respecting submarine curiosities.

For some years past I have kept several small aquaria, the inmates of which I obtain by dredging at sea, or by searching amongst rocks and stones on the shore: the study of my marine captives is a source of unfailing interest and amusement, and I feel sure that those of my readers who have had similar small tanks in their rooms, will agree with me in the assertion that the more one sees and knows of the inhabitants of the ocean, so much the more does one desire further information respecting them. If then such be the feelings excited by aquaria of very limited dimensions and modest pretensions, we may safely conclude that they will be increased a hundredfold by viewing the innumerable specimens of submarine animals displayed, under the most favourable circumstances, in the spacious tanks of the splendid building designed for public inspection at the "Queen of the Watering Places."

The site of this building is close to the Chain Pier, immediately below the cliff: it is protected from the waves by a strong sea-wall formed of concrete and Portland stone. The old road and promenade formerly leading to the Pier exist no longer, but a new road and promenade have been constructed south of the Aquarium.

The principal entrance to the main building is opposite the east side of the Albion Hotel, near the spot where, for so many years, stood the toll-house for the Chain Pier, well known in Brighton as "Ratty's toll-house;" upon referring to the ground plan of the Aquarium, my readers will observe a flight of steps leading to the building; these steps (twenty-three feet in width) are composed of granite, with spacious landings, and ornamental stone balustrades at the sides: descending these steps, the visitor finds himself in an open court, considerably below the new road-way. The dimensions of this entrance court are sixty feet by forty; it is arcaded on three sides by enriched terra-cotta arches, coupled columns, and entablature of Italian character in the same material; the columns have carved terra-cotta caps, red and buff, whilst the keystones of the arches consist of figures of mermaids; scriptural texts in mosaic letters being inserted in the frieze.

From this court the visitor passes on to the entrance hall of the Aquarium, an apartment about eighty feet by forty-five, with arcades on every side, composed of terra-cotta and ornamental brickwork; the vaulted roof of this hall is composed of stone ribs and coloured bricks, the centre being glazed for the admission of light. A corridor runs round the hall, the outer pillars of which are cast iron, elaborately worked; the inside, terra-cotta enriched columns of red and buff.

Facilities for refreshing the "inner man" are to be found at the north end of the hall, and close at hand is the entrance to a dining-hall—a commodious room, with a vaulted roof, supported in the same manner as the one just described, and lighted by plate-glass windows from the entrance court.

From the entrance hall the visitor proceeds to what may be termed the Aquarium proper, which consists, in the first place, of a corridor two hundred and twenty feet long and twenty-three wide, divided in the centre by handsome columns, constructed alternately of Bath stone, serpentine, and polished granite. On the caps of the columns are carved designs of shells, fish, sea-birds, and other marine subjects; while the corresponding corbels from the piers are enriched in a similar manner. From these columns and corbels spring the ribs of a handsome groined roof, the ribs being in Bath stone, and the filling in, between the groins, of brickwork; black and red, and red and buff alternately. On each side of the corridor are ranged the tanks for the exhibition of the specimens, varying in size from eleven feet six inches by twenty feet, to one hundred and three feet by forty. By re-

ferring to the plan it will be seen that the length of the corridor is broken by a central square, on the sides of which are the largest tanks in the building; the enclosed sketch representing the larger of the two. It may here be stated that the corridors throughout are divided into a series of bays of equal proportions; these divisions have Portland stone piers, plinths, and cills to receive the sheets of thick plate glass, which are secured to them with waterproof cement; exhibiting a front of water behind each division of glass, nine feet in length, and five feet six inches in height; and a glass roof covers the tanks, so as to admit a sufficiently strong light upon the inhabitants.

The central square has the groined vaulted roof on each side, forming a sort of cloister, whilst the middle is covered with an elaborately ornamented iron roof, part of the covering being antique coloured glass.

Passing through the principal corridor, which runs nearly from west to east, the visitor has access to another corridor at right angles to the one just described, or running directly north and south. There are no tanks in this part of the building, its chief use being a communication between the two main corridors.

But before giving any details of the second main corridor, I must not forget to mention the conservatory at the end of the one just described. Its extreme length is one hundred and sixty feet, and its width forty feet—the eastern end being designed to form a rockwork waterfall, with ledges, dropping well, small aquaria, ferns, &c. The conservatory proper is seventy feet long and twenty-three feet wide, the entrance facing the main corridor from the entrance hall. On each side are lofty arches forming arcades, having a tier of windows overlooking the terraces, and the roof is in cast iron, resembling the iron roof already mentioned.

The other main corridor, which is approached from the one running from north to south, corresponds in length to the conservatory, its width being the same as that of the principal one: it has twenty tanks, part for the exhibition of fresh-water specimens, and part for tropical specimens requiring a higher temperature. The conservatory has communication, on the south side, by means of the open arcade, to a part of this corridor, which is arranged for small table aquaria requiring minute inspection. The vaulting, lighting, glazing, and roofing of this corridor are similar to those of the first, and at the end are the engine rooms, for pumps and machinery, boilers, &c., also naturalists' rooms.

Underneath the corridors are large reservoirs for containing a supply of seawater; each reservoir communicates with the others, and with a well in the engine room. The pumps raise the water from this well to an upper reservoir, which has a pipe communication to *every* tank, to admit the flow of water at any required rate. A stand-pipe in each tank regulates the height for the water, by permitting its escape at the top, finding its way to the lower reservoir by earthenware pipes, to be re-used again—thus keeping the water in a perpetual flow from one end of the building to the other.

And now, having detailed the leading features of the *interior* of the Brighton

Aquarium with as little of the technical as possible, I must give a few particulars respecting the *exterior* of the building, trusting that the interest and patience of my readers are not quite exhausted.

Terrace promenades, decorated with stone balustrades, flowers, ferns, and shrubs, are constructed over the corridors, whilst seats and shady verandahs are provided for the accommodation of visitors. There are three flights of Portland stone steps leading to these terraces, one at either end of the building, and one in the centre, whilst from the new road next the sea is an entrance for invalid chairs by inclines, one leading to the entrance hall, and the other to the middle corridor.

Such, my readers, is a brief description of the architectural design, arrangements, and attractions of the Brighton Aquarium;

many other particulars could I give of this truly magnificent building, but as most of my readers will probably receive ocular demonstration thereof, a prolonged pen and ink description is unnecessary; therefore I will proceed to the consideration of a few of the denizens of the ocean, which, with innumerable other varieties, will be displayed in the richly ornamented and spacious tanks of the Brighton Aquarium.

But before passing on to the subject of Life Beneath the Waves, I wish to state that this little book is not intended to be what is called a scientific work; its object is simply to draw attention to the inhabitants of the sea by short descriptions of various specimens, and their most striking characteristics, and to thereby produce desires for further information in the minds of those who were previously totally indifferent to the fact that, hidden away in the depths of the ocean, live tribes of creatures of the most exquisite beauty and wondrous organizations.

So multifarious are these submarine denizens—so interesting, marvellous, and attractive their various idiosyncrasies—that it is rather a difficult matter to decide which to select for consideration in a little book of this description.

And again, on the one hand, if I indulge in minute details, the charge of trespassing on other people's grounds may be laid to my account; and on the other, if but few particulars are given concerning the submarine specimens selected, I may be accused of ignorance and inability.

So you see, I am, as Brother Jonathan would observe, "rather in a fix;" but as "fixes" do not, as a rule, "become small by degrees and beautifully less" when unduly considered and brooded over, I will no longer dwell upon the one now in hand, but decide to limit myself, in the following pages, to the description and delineation of some of the submarine animals which have from time to time occupied my own small Aquaria.

For some few years I have contributed short papers on these subjects to the well known and popular magazine the *Chatterbox:* with the kind permission of the editor, parts of these contributions are now reprinted in this little work.

PART II.

The tanks wherein anemones are displayed will doubtless be some of the most picturesque and attractive in the Brighton Aquarium, for although when out of water the anemone is a shrivelled mass, unsightly to the eye, and flabby and unpleasant to the touch, its appearance when *in* water fully justifies its appellation of "Flower of the Sea."

There are, I believe, twenty known kinds of British Anemones, some of them greatly outrivalling others in point of beauty, as well as in their resemblance to flowers.

The Dahlia Anemone is a beautiful

species, and bears a striking resemblance to the flower from which it takes its name. It belongs to the group scientifically termed Bunodes Crassicornis; a specimen now in my possession measures seven inches across when fully expanded. The colour of its rough base or disc is orange, with white protuberances, its centre rosy-pink, streaked with dark crimson, and softening into white at its mouth—the outer rows of its transparent tentacles are also white, with crimson marks—the inner rows faint pink, also streaked with crimson.

This anemone is extremely greedy and voracious, shrimp after shrimp being caught in its extended tentacles, and ere one victim disappears in the Dahlia's mouth, another is often captured, and gradually drawn into the same place.

One of the peculiarities of this anemone

Pl. 1.

W. West & Co. imp.

Dahlia Anemone.

is the variety of shapes it continually assumes: when fully expanded, and one is admiring the regularity of its tentacles, and its general perfection of form, suddenly, and without any apparent reason, some of its transparent feelers become contracted, and are withdrawn into its disc, thus causing the creature to appear what is called "lop-sided;" at another time, instead of altering the arrangement of its tentacles, its mouth and interior are protruded in such an extraordinary manner, that the anemone seems to be actuated by the desire of turning itself inside out, to use a common expression; and sometimes it suddenly closes itself up altogether, and instead of reminding one of a beautiful Dahlia, it looks more like an ill-shaped and unwholesome orange.

Other anemones in the same tank with

the Dahlia belong to the class scientifically denominated Actinia mesembryanthemum. The bodies or discs of this species are quite smooth, and devoid of the rough excrescences which distinguish the bases of the Crassicornis tribe: Smooth Anemones vary very much in colour, some of them being entirely crimson, green, or brown, whilst others are ornamented with stripes or spots.

The Strawberry Anemone is a pretty, though common, specimen; thus called because, when closed, its colours and general appearance resemble that delicious fruit. When the Strawberry is opened, and its tentacles fully displayed, a row of bright blue dots may generally be seen at its edge: these blue dots may also be observed in most of the anemones belonging to this class.

Smooth Anemones may be found of all

Smooth Anemones.

sizes as well as colours, but I believe they never attain the proportions of the Dahlia just described—the largest in my tank at the present time is about four inches across when fully expanded; the smallest, about the size of a pin's head: the number of these tiny creatures continually increases, and very pretty do they look, dotted about the aquarium either on the pebbles or seaweed, with their little arms widely extended.

The tentacles of the Strawberry, and other smooth anemones are very much longer and thinner in proportion than those of the Crassicornis family; and although perhaps the Smooth Anemone is not quite so voracious as the Dahlia, it is by no means of an abstemious nature, for even the tiniest specimen will readily seize and devour bits of meat or fish, as well as living shrimps, and any other of their companions in captivity which seem good to them, and which are unwary enough to venture within reach of their murderous tentacles.

It is, perhaps, not generally known that various kinds of worms inhabit the sea, as well as those which live in the earth; and that whilst earth-worms with their dull colours, and wriggling movements are not calculated to inspire interest and admiration, most sea-worms, or Annelids as they are properly called, are extremely pretty and attractive objects.

Some of the most beautiful of this group of the denizens of the sea are those denominated Serpulæ, and Sabellæ; these worms inhabit tubes, something like the stem of a clay pipe in shape and colour, and more or less bent and twisted. The inhabitant of each tube can push out and draw back, at its pleasure, a bunch of feathery tufts, extremely delicate in texture, and most brilliant and exquisite in colour. These feathery tufts are the gills of the Annelid. They vary very much in hue, some being scarlet, some grey, green, blue or brown. Others, again, are orange, dotted with black, or light yellow, edged with white; in short, the various colours are so numerous, that when a group of bright tufts is displayed at one time, the effect is almost as gay as that produced by flowers in a conservatory. At the first glance no very particular difference is to be observed between Serpulæ and Sabellæ, but a closer examination discovers distinct characteristics.

Upon looking at the illustration of the Serpulæ, something like a little cork on a stem will be noticed amongst its plumes; when the Annelid draws back its feathery tufts, this little cork goes into the shell *last*, and acts as a stopper to the opening of the tube; and when the gills are protruded, the cork makes its appearance first, as a matter of course. Every Serpula has two horns or antennæ, the one being ornamented with this tiny cork—the other being shorter, and, as a rule, devoid of any such "terminal cone" (as a well-known naturalist denominates the stopper). But I have occasionally found Serpulæ with a stopper on *each* of the antennæ, and upon examining the movements of these specimens through a magnifying glass, I observed that when the feathery plumes were withdrawn, one of the little corks followed slightly behind the other; and when the gills were again protruded, the

Sabellæ.

antennæ appeared first, in the same order in which they had disappeared; and as they issued from the tube, closely followed by the beautiful plumes, with a jerking movement they righted themselves side by side. Upon looking down the tube when closed, I could only perceive one of the stoppers, and as it fitted exactly into the tube, the natural conclusion is, that the two "terminal cones" act like two doors, thereby rendering the inmate of the shell doubly safe from injury. The shape of these little stoppers is not always the same, for some are pointed at the top, and bear a resemblance in shape to a tiny mushroom. The colour of these Serpulæ is generally brown, drab, or black.

One of the principal differences between Serpulæ and Sabellæ is, that although the feathery plumes of the latter are more delicate and fragile in appearance than those of the former, they have no protection in the way of a stopper when hidden in their tubes. The elegance, lightness, and gracefulness of these gills of the Sabellæ are indeed beyond description; like the Serpulæ, they vary in colour. Those I have sketched were delicate gold, with brilliant scarlet dots; some are pale green, streaked with black, and when fully opened, and waving backwards and forwards in the water, they are most exquisite and beautiful objects. Another distinction between the Annelids of which I am writing is, that whilst the Serpulæ are usually found in groups adhering to broken shells or pieces of rock, the Sabellæ I have drawn are, as a rule, seldom to be obtained even in pairs. And whilst the tubes of these Sabellæ adhere but partially to their resting-places, three-fourths of their shells generally being quite independent of any support, Serpulæ, on the other hand, adhere closely to rocks or shells, and only a small portion of their tubes stands aloof.

Both kinds are excessively shy, and very easily frightened, the slightest noise or movement of the aquarium causes them to retreat into their tubes, and very quiet must the observer remain if he desires to see the gay plumes come forth again.

And yet, strange to say, when Serpulæ adhere to a shell whereof a Hermit-crab has taken possession, the creature's rapid and abrupt movements do not alarm the Annelids, for they extend their gay plumes to the utmost width, as the crab hurries about the aquarium in a manner by no means remarkable for its lack of noise.

But if a finger is laid on the glass, or the breath of the beholder ruffle the water, the tufts disappear in a moment. Serpulæ and Sabellæ appear to possess hardy natures, for I kept the specimens I have drawn, for more than twelve months; and at the end of that period they were evidently quite healthy, for their plumes were as brilliant as at first. I then left that part of the country, and gave away my marine pets, greatly to my disinclination I must confess.

Whilst dredging at sea one day, at a distance of twelve or fourteen miles from the shore, the first thing which attracted attention, as the contents of the dredging-net were emptied on the deck, was a piece of rock about six inches in length, and four in depth, completely covered with Serpulæ tubes of various dimensions; and upon placing it in an aquarium after returning home, it soon became evident that other Annelids besides Serpulæ had taken possession of the rock.

The first intimation of the presence of these worms was a bunch of long, thin, buff-coloured feelers or tentacles, ornamented by a short thick crimson fringe, hanging down from the piece of rock. These feelers were continually in motion, elongating and shortening themselves every instant. And as I examined them through a magnifying glass, the bunch moved forwards, and in a few moments a salmon-coloured worm had displayed itself, to the length of about an inch and a half.

Thinking that in all probability the Annelid was searching for food, I dropped a small piece of meat near the perpetually moving tentacles, and in an instant it was enveloped by the numerous feelers, which twined and untwined themselves round it with great rapidity, the tentacles appearing to act for the worm as our teeth act for us, with the exception of theirs being *outside* instead of *inside* the mouth, for the piece of meat was gradually softened and divided into very small particles, and drawn up beneath the crimson fringe.

These Annelids lived but a week or two; they came out of their houses in the rock, and lay extended amongst the pebbles some hours before they died, the restless tentacles gradually becoming quieter, until at length they ceased to move at all.

Serpulæ and Sabellæ, and many other worms which inhabit tubes, forsake their shells in this way before death; their brilliant hues generally become fainter

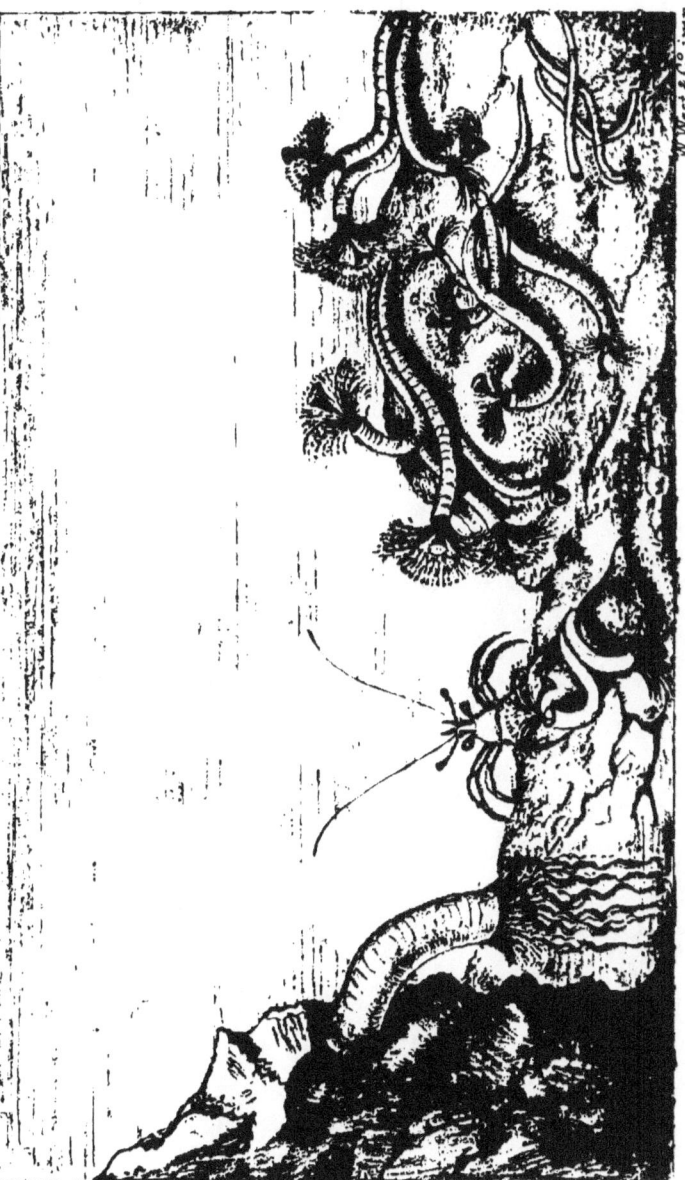

Sea worm. Serpulæ

and fainter as their life ebbs away, and fade entirely after death.

Besides the Annelids just described, numerous specimens of those which do not inhabit tubes have been captured in our dredging net; the colouring of these creatures is also extremely brilliant and varied, and their movements are both active and graceful, and very pretty do they look as they dart amongst the stones and seaweed of an aquarium. The one I have drawn was scarlet, with white marks on its body and a white line on its back, reaching from its head to its tail. The sketch is the same size as was the worm itself; and I may as well observe that all the illustrations of marine creatures in this little book, are of the same dimensions as the originals, unless mention be made to the contrary, and that, with one or two trifling exceptions, they are drawn from specimens which have inhabited my own aquaria.

A broad back, strong claws, tiny eyes, and side-long walk are familiar associations in our minds with respect to the common or edible crab; but the fact is not perhaps equally familiar, that when this "dainty morsel" is, so to speak, in his infancy, his outward form bears no resemblance to that which it attains when he arrives at years of maturity.

Indeed so very different are the Zoea, or young crabs, from the old ones, that at one time they were supposed to be a distinct race of creatures. The transformations they undergo before attaining the shape familiar to us all are very extraordinary, and well worthy of description.

In its first stage of existence the young crab has, says a writer on this subject, "a helmet-shaped head, terminating behind in a long horn, and furnished in front with a pair of huge sessile eyes, and it moves through the water by means of its long swimming tail. After the first change of skin the body assumes something like its permanent shape; the claws are developed, and the legs resemble those of the crab; but the change is still incomplete, for the tail is still long, and furnished with false feet like that of the lobster. The swimming habit has not yet been laid aside. At the next stage, while the little creature is still about the eighth of an inch in diameter, the crab-form is completed, the abdomen folding in under the carapace. All the subsequent changes are merely changes of coat, consequent on the growth of the now complete animal. In these several metamorphoses we see portrayed in succession the peculiarities of three different types, one rising above the other in structure. In the first stage the crab resembles one of the least perfect Crustacea, such as the Water-flea; in the next, it assumes the aspect of the lobster; and finally puts on the form of the most perfect animals of the class."

Another curious circumstance in connexion with the edible, and other crabs, is that, contrary to the general course of nature, the shell does not expand in proportion to the creature's growth, consequently he is obliged from time to time to cast it off.

It is very interesting to watch the crab as he rids himself of the coat which has become too small for him. I have seen the whole proceeding many times. One crab in

Scarlet-Worm.

Pl. 6

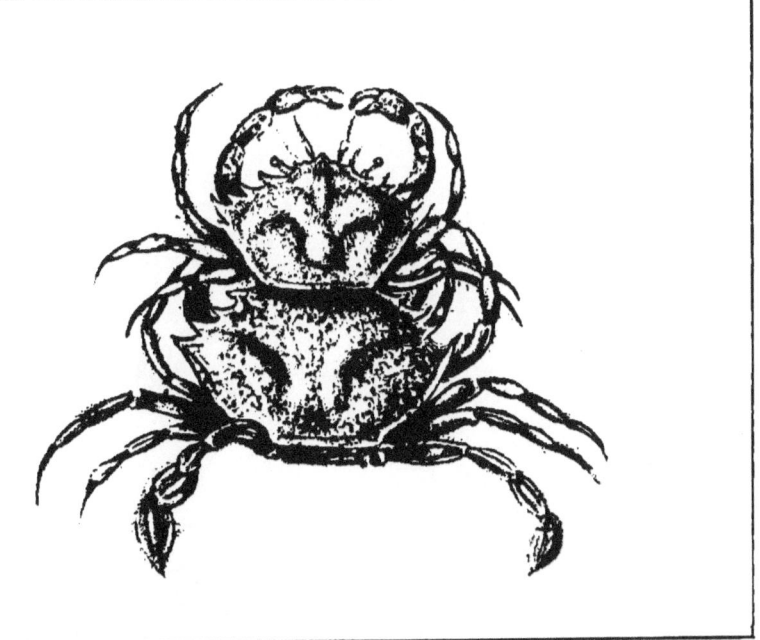

Crab changing its shell.

particular which lived in my aquarium, grew very quickly, and, if I remember rightly, threw off his coat six times in less than a year. At length he became so large that he continually made his escape from the glass, generally selecting the night for that purpose, and one of my first acts, upon entering the room in the morning, was to search for the truant and put him back in his home. His propensity for persisting in thus roaming by night terminated fatally, for he was unfortunately crushed to death beneath the fender. I was sorry for his untimely decease, for he had long been in my possession, and always appeared to be in an excellent state of health and spirits; we had moreover become quite upon friendly terms, and for many months before his death he had ceased to exhibit any symptoms of alarm when I approached the aquarium; and when a small stick was inserted, with a modicum of food attached, the moment the piece of wood touched the water, he would scramble towards it with great rapidity, and eagerly snatching off the bit of raw mussel or other fish, dart away to the nearest shelter of rock or seaweed, and devour the morsel with every appearance of relish and enjoyment.

The first sign that a crab is about to cast off his shell is generally an extra amount of restlessness—round and round the aquarium, backwards and forwards he crawls, till he finds what he considers a sufficiently retired spot for the purpose of making his strange toilet. He then remains stationary, and the shell begins to move forwards very slowly, the crab's claws being slightly elevated at first, and his legs extended. On moves the coat, very gradually, but perceptibly, and as it gets more and more forward, we perceive that the crab is gently doubling up his legs and lowering his claws, till at length the shell which has hitherto protected him protects him no longer, but falls from him, not broken into pieces, but *perfect*, even to the covering of the eyes and antennæ or horns. The marks on the crab are precisely the same in shape as those on the discarded coat, but larger, and the colouring very much more brilliant.

So greatly is the crab now increased in size, that our first thought upon viewing him without his discarded coat, is surprise that he could ever have worn it at all; and this idea seems generally to occur to the crab also, for his frequent custom, on getting rid of his shell, is to seize it with his claws, and after examining it for a moment with apparent astonishment, he pushes it contemptuously away, and hastens to find some snug hole or obscure corner wherein he can hide himself, till he can again issue forth in all the glory and comfort of a new suit of armour.

His present skin is now very soft and tender, and by no means calculated to act as a defence against the attacks of his companions—those of his own kind being especially ready to take every advantage of his weak and forlorn condition. Upon one occasion I noticed a tiny crab crouching down under a stone, close to the spot where another crab, twice his own size, was occupied in casting off his shell; whether the little one owed the larger one a grudge, or not, I cannot say, but no sooner was the latter rendered defenceless by the loss of his armour, than the little wretch darted at him, tore off one of his legs, and actually proceeded to *eat* it.

In order to save the poor denuded crab from further insults and cannibalistic attacks, I placed him in another aquarium, where there were none of his own species, and there he remained until his shell became hard again, and he could fight his own battles.

Judging by their behaviour in an aquarium, crabs of all kinds are, as a rule, extremely quarrelsome and pugnacious; they live in a state of perpetual warfare; and their custom of pulling off each other's limbs, and demolishing the same with the greatest gusto, is of very frequent occurrence. Strange to say, the loss of one or more legs does not seem to interfere much with the comfort or well-being of the victim, for he scuttles about as quickly as ever on his remaining members, and takes the first opportunity of playing the same trick on his comrades.

It is worthy of remark that new legs soon appear in the place of the old ones; and that crabs, lobsters, and other Crustacea have the strange power of breaking off their legs and claws at will, when alarmed, or enraged: this circumstance accounts for the difference in size often seen in the large claws of crabs and lobsters, or for the absence of one claw altogether, or of one or more of their legs.

There are a great many different kinds of crabs: some can be obtained only by dredging at sea, some may be found amongst the rocks on the shore, and others are seen crawling about on the sand when the tide is out.

Amongst those to be found by dredging in deep water, is the crab known by the

Hermit-Crab.   Angled-Crab.

name of the Angled-crab; its chief peculiarities are its immensely long claws, or arms, its long-stalked eyes, and its having on its carapace or shell a curiously distorted impression of a human countenance. The colouring of the Angled-crab is very pretty and somewhat varied. The smaller specimens are generally of a delicate pink hue, merging into white; in larger specimens, the arms, legs, upper part of the back, and the stalks of the eyes are salmon-colour, the lower part of the carapace being much lighter, whilst the moveable joint of the large claws is densely black, also the shining eyes.

The disposition of the Angled-crab appears to be more amiable than that of many of his relations; I have continually had specimens in my aquaria, and have never seen them attempt to quarrel, fight, or eat their companions; on the contrary, they allow the smallest crabs to crawl quite close, and even *over* them without displaying any annoyance or wrath, and if too much advantage is taken of their good nature and peaceable disposition, they gently push the offenders away, or betake themselves to a more secure place of refuge.

The Nut-crab is another kind to be obtained only by dredging at sea, and a quaint-looking little creature it is, with its particularly short legs, plump arms, and curiously shaped shell, whereon, as with the Angled, a strange resemblance to a human face is depicted. Very sluggish in its movements is this little Crustacean; very shy and retiring its disposition. It does not stir about much when in captivity, but appears perfectly contented to squat amongst the stones, or to bury itself in the sand; a desire of attracting observation is certainly not one of its characteristics, but, after a time, it will venture from its hiding-place if offered food, which it will readily seize and demolish.

Its colour is generally light brown, or dull white, and in some specimens the plump little claws and legs are tinged with red.

The distinguishing features of Spider-crabs are totally different from those of our compact-looking little friend the Nut—very long thin legs, triangular-shaped bodies, with a sort of snout in front, being their most striking characteristics.

There are many varieties of this class—some of them being far more like spiders than others. The body and legs of the one depicted with the Nut are semi-transparent, and ornamented with fine silky hairs, its colour is dark brown, with white marks on the body, and it moves gracefully along with a languid floating motion. It is a very pretty object in an aquarium, but I have not been able to keep it alive for more than a few months. The other Spider-crab sketched is a much commoner species, and is not nearly so light and graceful, its body being opaque, and hard, and covered with sharp spines and knobs, and generally profusely adorned with sea-weeds, which sometimes flourish so luxuriantly that the shape of the crab is scarcely perceptible; and very curious does he look as he crawls slowly through clear water, with his burden of variegated seaweeds waving around. The legs are long and slender, but the arms are strong and stout, and both legs and arms are

Spider – Crab.

covered with thick short hair; its movements are sluggish, and its habits retiring.

Another crab remarkable for the extraordinary length of its arms is that known by the name of the Helmet-crab, which title is bestowed because the body is something like an old-fashioned helmet in shape. It is rather a rare species, and seldom to be obtained alive without dredging at sea; its arms are indeed enormously long, and when stretched out, they look very awkward and disproportionate: the upper part of the body and legs is fawn colour, the under side white, and on its back (as on that of the Angled-crab, Nut, and some other kinds) a distorted likeness to the human face may often be observed.

And now I must mention another class of crabs, totally different, in one respect at least, from those already described. Edible crabs, and many of their relations, can crawl with great rapidity at the bottom of the sea, or on the shore, but they are quite unable to swim. But when dredging in deep water, various specimens of swimming crabs may be obtained, and they form very interesting additions to an aquarium, their movements being so different from those of other species.

The bodies of these crabs are much broader than they are long, and much wider in front than behind, and their eyes are placed a considerable distance apart; their chief peculiarity, however, consists in their hindermost legs, which are broad and flat at the end, and are used by their owners as oars or paddles, thereby enabling them to swim or float with the utmost ease. But the fact of these crabs being able to " paddle their own canoes" in such a truly *natural* manner, does not debar them from pedestrian movements, for swimming crabs can crawl, and scramble *over* rocks and stones, as easily and rapidly as they can swim *above* them. A beautiful member of this family is the Velvet-crab, so called by reason of its body and limbs being thickly covered with hair, as fine as silk and as soft as velvet; its colours are golden brown, and blue, and it is a most attractive object in an aquarium. Its disposition, however, is not as worthy of admiration as its personal beauty, for the Velvet-crab is particularly fierce and quarrelsome, and its cannibalistic propensities are decidedly objectionable in tanks of small dimensions.

Before leaving the subject of crabs, I must touch upon one other kind, for it is a particularly amusing inmate of an aquarium, and that is, the Hermit, or Soft-tailed crab.

In shape he is totally different to any of those previously mentioned; in fact, the Hermit bears a greater resemblance to a lobster than a crab, his body being long and thin, instead of flat and broad; and he, moreover, possesses a tail, which, strange to say, is quite soft, and devoid of the protection of shelly armour, which covers the fore part of his body, arms, and legs.

In consequence of this deficiency of his "latter end," the Hermit takes up his abode in a shell, in which he fixes himself very firmly by means of a pair of pincers at the end of his tail, so firmly indeed, that it is a difficult matter to pull him therefrom without injury. As a rule, one of the Hermit's arms is larger than the other; his horns are of a considerable length; his eyes are

Pl.10.

Helmet-Crab.

Swimming-Crab.

mounted on long stalks, and very inquisitive and impudent does he look, as he peeps out of his place of refuge; he moves about very quickly, and does not seem in the slightest degree inconvenienced by the weight of his house on his back.

The Hermit may be found in all kinds of shells, and when he grows too large for, or becomes tired of living in one shell, he changes it for another. When in captivity, the Hermit is often very restless, and will continually change his residence, if suitable shells be given him. The way he conducts himself during the transit is very amusing —I have often watched his comical evolutions as he forsakes one abode for another.

As soon as an empty shell is dropped into the aquarium within the Hermit's sight, he darts eagerly at it, seizes it with his claws, and after dragging it away to some retired spot, proceeds to examine it thoroughly, both inside and out, thrusting his arms and legs *into* it, and *over* it, as if anxious to discover whether its dimensions are likely to suit him. If not satisfied with his examination, the Hermit pushes the shell away; but if he considers it worthy of a trial, he suddenly draws himself out of his old house, and jerks his tail into the new one with surprising rapidity; if the new habitation prove comfortable, the Hermit remains therein, but if not, he returns to his old quarters as eagerly as he quitted them.

If the Hermit forsakes one shell without making an attempt to enter another, it is generally a sign that he is ill, and likely to die. Occasionally his companions will allow him to draw his last breath without molestation, but more frequently the poor creature is torn in pieces, and devoured within a few minutes of his quitting his place of refuge.

Sometimes, instead of desiring to ensconce himself in an empty shell, the Hermit takes a fancy for one wherein its legitimate owner is still living, and I have read that on such occasions he will "seize the animal with his claws, and having devoured its flesh, take the shell for his own use," thus forming a striking illustration of the saying that "Might is right."

The appellation of Soldier-crab is also bestowed upon the Hermit, in consequence of his excessive pugnacity, and love of fighting. He soon becomes reconciled to captivity, and seems perfectly happy scrambling over the stones in pursuit of his companions, or resting under seaweed when tired of warfare.

Many other species of crabs besides those just mentioned, have, from time to time, inhabited my aquaria; and so numerous are the varieties to be found in the ocean, that volumes might, (and have been) written respecting them: but as the subject of crabs, however interesting, is not the only one now to be considered, I must refrain from further dilatation thereupon, and invite the attention of my readers to other groups of marine curiosities.

Few people walk along the seashore without picking up mutilated shells of various descriptions, but these broken shells, pretty as they may be deemed, give but a very inadequate idea of the beauties of living, perfect specimens.

The shells denominated Pectens are amongst the most beautiful of British

Pl. II

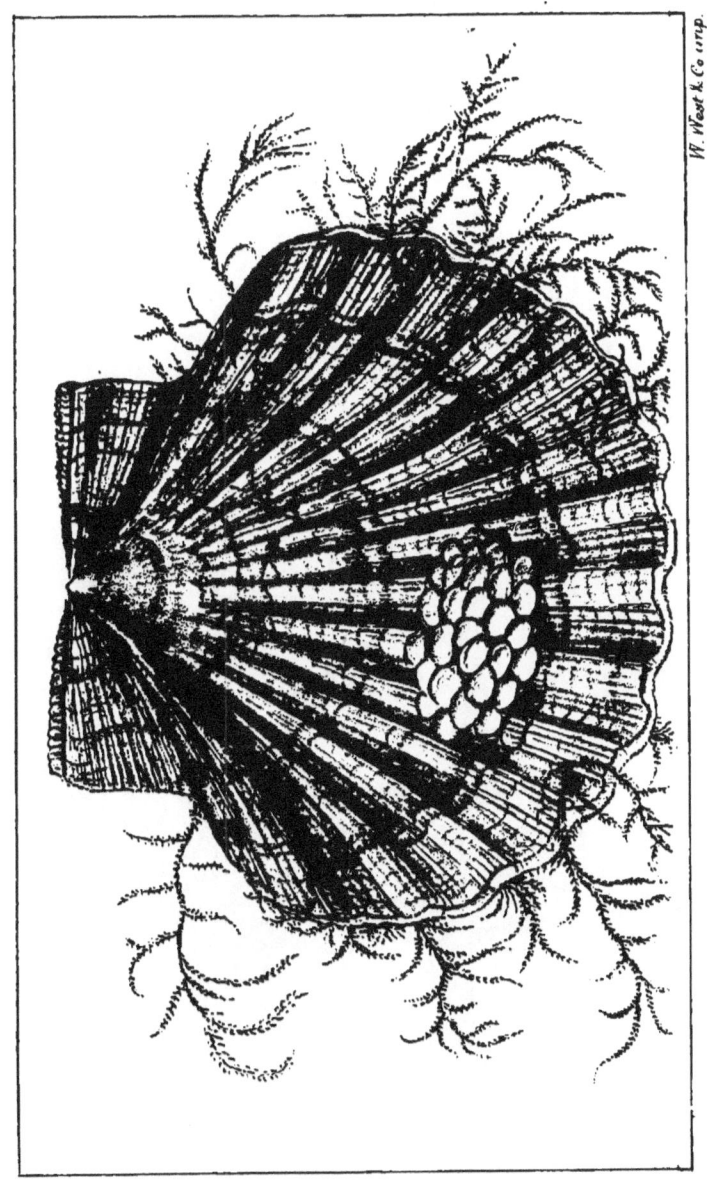

Pecten or "Escallop."

groups. I have obtained a great number at different times with the dredge, and so various were they in size, colour, and pattern, that it was a difficult matter to decide which were the most worthy of admiration, which to keep, and which to throw away.

The colouring of Pectens dredged up at sea is very gay; some specimens are bright yellow, marked with crimson; some, light brown, or red, dotted over with black; others, a delicate pink, with lines of dark rose colour; others again, are all white, or orange, or pink, as the case may be, with no marks at all; and some are rich chestnut or fawn colour, ornamented with sharp white points. It is, in short, quite impossible to describe half the exquisite tints and shades of these beautiful Sea-butterflies, as Pectens are also called, owing to their peculiar mode of moving about: for instead of crawling from place to place, in the same way as most other shells, Pectens spring lightly through the water with a series of leaps, caused by quickly opening and shutting their valves, something after the manner of a butterfly's wings.

The best time to see a Pecten to perfection is after it has been in an aquarium (or any other vessel filled with sea-water), for two or three hours; its shell will then generally be open about half, or a quarter of an inch, and we find that beautiful as is the exterior of the Sea-butterfly, the live inmate of the shell is, if possible, still more capable of inspiring us with admiration and interest. When open as just mentioned, a brightly coloured narrow band is visible on either half of the shell, and on the edge of each band is a row of tiny dots of the most brilliant hues, glittering and flashing as brightly as jewels; these dots are supposed to be the eyes of the Pecten; a double row of white tentacles is to be seen on each band, those attached to the lower, or inside edge, being shorter and finer than those at the top, close by the shell; these feelers are continually in motion, waving to and fro in the most light and graceful manner.

The Sea-butterfly fixes itself to rocks or stones by means of a bunch of fine silky threads which it has the power of spinning; they issue from the top or hinge of the shell, and are properly termed " Byssus."

The common name of these beautiful bivalves is "Scallops," or as it is properly spelt "Escallops;" and to some people they are known merely as an article of food; very large quantities are sold for this purpose, and they are stewed, scalloped, or curried, according to taste. The large Pecten, with the seaweed attached, was amongst some others which I bought off a fisherman's barrow, the white bunch on the shell is the spawn of the common whelk. In "Chambers's Encyclopædia," we are told that "Pecten Jacobæus, a native of the Mediterranean, is the scallop-shell which pilgrims were accustomed to wear in front of their hats in token of their having visited the shrine of St. James at Compostella. It attains a size of about four inches long and five inches broad."

It is quite impossible for a pencil drawing to give anything like an adequate notion of the beautiful hues of Pectens, and it would moreover be very difficult for a painting to do ample justice to the extreme brilliancy or delicate transparency of the colours of the lovely Sea-butterflies.

The Pelican's-foot is a very pretty and curiously shaped shell; its chief peculiarity

Pl.12

W. West & Co. imp.

Pecten varius, Dogwinkle. Pelican's-Foot.

consists in its having a large lip ending in several points, and bearing a resemblance to the webbed foot of the Pelican (whence its name).

The size of this mouth or lip depends upon the age of the shell; when young, the lip is small, but when the shell arrives at years of maturity, the lip becomes disproportionately large.

Numerous marks will be seen on the shell, upon looking at the illustration: these marks are prominent, and are generally reddish-brown, on a surface of the same hue, but considerably lighter. The inside of the lip is smooth, and of a faint buff, or salmon colour.

The shells called Trochus, or Top-shells, are very numerous on some of our shores, and a large quantity of different kinds and sizes may be picked up in the course of a few hours; but as all these shells are more or less chipped, and otherwise damaged, they are of no value in the eyes of a collector.

Those procured by dredging are almost as gaily tinted as the Sea-butterflies; the groundwork of some specimens being yellow, with marks of bright crimson; others, fawn colour, marked with delicate pink. If muriatic acid be applied to these shells the outer surface will be destroyed, and a coating of delicate pearly whiteness, shot with crimson, green, and gold will be disclosed; in this state they are frequently made into breast-pins and other ornaments.

Chitons, or Mail shells, belong to the class of Sea-slugs; their bodies are well protected by a shell arranged in plates, something like ancient armour, from which circumstance they are called Mail-shells; the plates lap over each other, and are

Pl. 13.

Trochus or Top-Shells.   Naticas.

W. West & Co. imp.

joined at the bottom to a sort of band which goes quite round the animal. If a Chiton is removed from its resting-place and held in the hand, it generally rolls itself up into a ball, after the manner of a woodlouse ; indeed, in some respects the outward appearance of a small Chiton resembles that of the insect just mentioned. Mail-shells move slowly along by means of a foot attached to their stomachs, and of the same length as the creatures themselves.

British Chitons are much smaller than foreign ones ; I have seen engravings of the latter measuring three or four inches in length, and two in breadth, whilst the Chiton marmoreus, the largest British species, is only an inch and a half long, and about seven-eighths of an inch broad. This kind is, I believe, never found on our southern coasts, but belongs especially to the north and north-east. The particular specimen I have copied (which was kindly lent to me for that purpose) was obtained off the Irish coast; the colouring is dark brown, with marks of olive green and chestnut.

The smaller Chiton represented is the commonest British species, and may be found amongst rocks and stones at low water. I have various specimens of this kind in my aquarium, but their numbers continually diminish, thanks to the crabs, who never fail to take advantage of any unwonted "absence of mind" on the part of the poor Chitons, and to make their shells remarkable for "absence of body."

Amongst the broken shells cast up by the waves on the shore, the Natica or Sea-snail is common enough; the Necklace Natica is the largest of this class. When

Chitons or Mail-shells. Banded Venus.

alive it has a smooth shining surface, usually yellow or brown, with streaks of deep crimson on its spiral rings; we are told that it is called the Necklace Natica on account of "the curious ribbon-like form in which its eggs are laid, somewhat resembling a broad necklace of pearls."

The chief peculiarity of the Necklace Natica is the extraordinary way in which the animal spreads its body over its shell when crawling about from place to place.

The lesser Sea-snail of the two illustrations is the prettiest shell; like the one just described, its surface is smooth and polished; its colour bluish-white, with rows of violet; the streaks are arranged in rows over the entire shell, not merely at the top like those on the Necklace Natica.

The pretty little shells called Cowries, or Tokens are very plentiful on some of our coasts: like Sea-snails, they are also remarkable for almost concealing their shells with their bodies when moving about; they are generally white, tinged with pink, brown, or grey, and sometimes spotted with black.

The Purpura, or Dog-winkle, as it is commonly called, is not at all a rare shell, for living specimens may be found in abundance amongst the rocks when the tide is out, and dead shells may be found in equal profusion on the beach.

But in spite of its being what is termed a "common shell," the Dog-winkle is well worthy of our notice; it was highly prized by the ancients on account of its having some yellowish matter in a sort of bag behind its head, from which a purple dye was made.

I have read that the inhabitants of Tyre

were famous for wearing purple cloth, the dye of which was procured from shells B.C. 2112, and that it was also worn in Greece B.C. 559, and that after having been lost for ages, the method of producing the Tyrian purple was discovered again in the seventeenth century.

The Venus shells are some of the most attractive of British groups; not only are their colours very pleasing, but their valves are, moreover, ornamented with raised lines or bands running parallel to the edges of the shells. The specimen drawn is called the Banded Venus; its surface is covered with broad bands of a delicate pink, very much raised, and separated from each other by an indented line—from the hinge to the margin are three dark-purple streaks, narrow at the top, and gradually increasing in width as they approach the edge of the valve. In spite of the delicate hues of the Banded Venus, its appearance is by no means fragile; on the contrary, it looks, as it is, very strong, the valves being solid and thick.

The Tapestry or Carpet shells belong to this group: this name is given because the arrangement of colour is thought to resemble that of the needlework so much in vogue amongst ladies many years ago—*tapes*, their Latin denomination, signifying tapestry.

These shells live in mud, as has frequently been testified when brought up by the dredge; indeed, the contents of the net have on some occasions been so extremely unprepossessing when emptied on the deck, owing to the mud with which they were enveloped, that it required both time and patience to render them fit for ex-

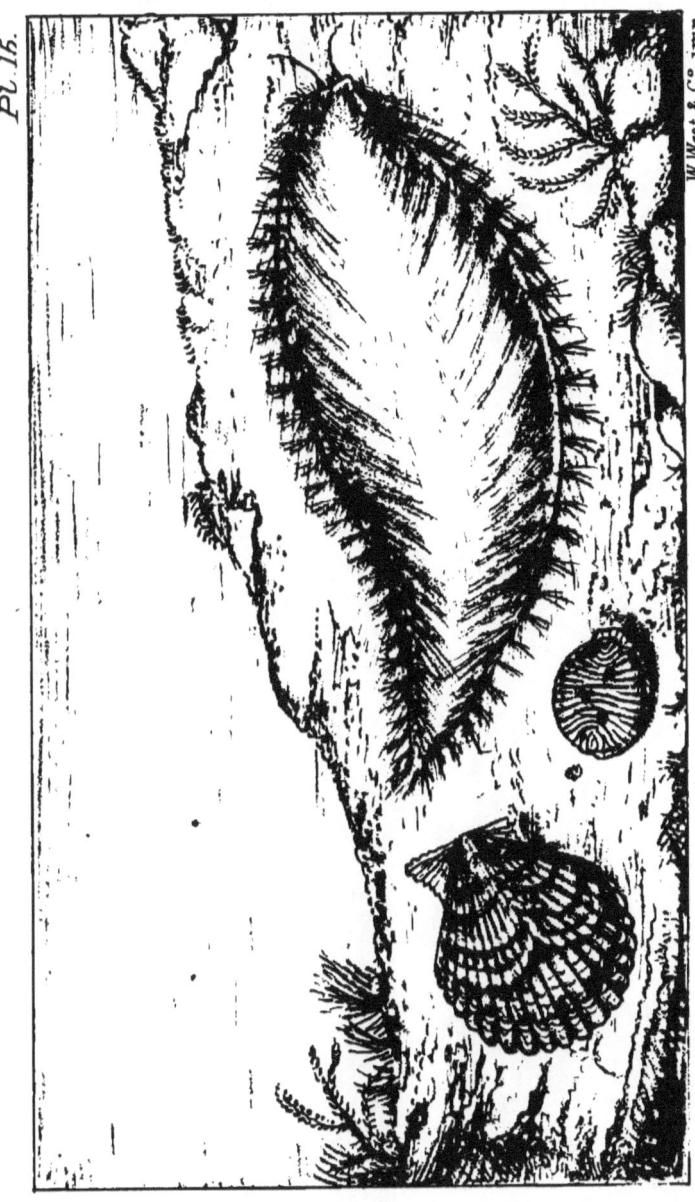

Pecten. Cowry Sea-Mouse.

amination by repeated applications of sea-water.

Another inhabitant of muddy ground is the Sea-mouse—a creature sufficiently ugly and uninteresting when covered with its "native soil," but remarkable nevertheless for its extreme brilliancy and variety of colour. The body of the Sea-mouse is brownish-grey, its sides are thickly covered with fine soft hair of the most gorgeous hues, exquisitely glossy in appearance, and continually changing in tints, giving one the idea of what is called a "shot silk dress." Amongst this beautiful hair are rows of strong bristles or spikes, which the Sea-mouse withdraws and pushes out at pleasure, and in all probability uses as weapons of defence.

As Mr. Wood remarks, in his "Common

Objects of the Seashore," "it would be a most unpleasant circumstance if the creature were to wound itself with its own weapons. In order therefore to obviate this difficulty, each spear or bristle is furnished with a double sheath, which closes when it is retracted into the body, and opens again when protruded. It is hardly possible to conceive a more wonderful structure in the whole of the animal kingdom."

The Sea-mouse is also provided with a number of stiff feet, which are also capable of retraction and protrusion.

Still more unprepossessing at first sight than the mud-covered Sea-mouse is a certain soft, dirty white or yellow substance, found adhering to shells and pieces of rock, and owning a name as unattractive as its appearance. Which name, however, is by no means inappropriate; the shape of this unsightly mass frequently bearing a striking resemblance to dead men's fingers, or dead men's toes.

But if a piece of the Alcyonium digitatum (as "Dead Men's Fingers" are scientifically called) be placed in water, we soon become aware that it possesses various beauties and attractions, and that, instead of being an ugly, inanimate lump, it is in reality a mass of the living creatures called Polypes.

If in a healthy state when placed in sea-water, the surface of the specimen of Dead Men's Fingers quickly becomes covered with these little creatures, which push themselves out from the unattractive lump, and entirely change its aspect. The tough mass is the polypidom, or general body of these polypes; and, strange to say, each polype assists to support its companions;

for the nourishment received by each member of the family group passes into the polypidom, and thus benefits not only the receiver, but *all* his companions. When facts such as these are taken into consideration, the Alcyonium is not to be despised or deemed uninteresting on account of its frequent unattractive shape and appearance.

Another submarine creature, apparently devoid of attractions or pretensions to beauty when out of water, is the Starfish; and as it lies on the beach, in a limp and lifeless state, it is certainly not a very fascinating object; but a living specimen of even the commonest kind is well worthy of observation when placed in water.

Starfish are provided with an immense number of sucker feet, which they stretch out and retract at will, and by means of

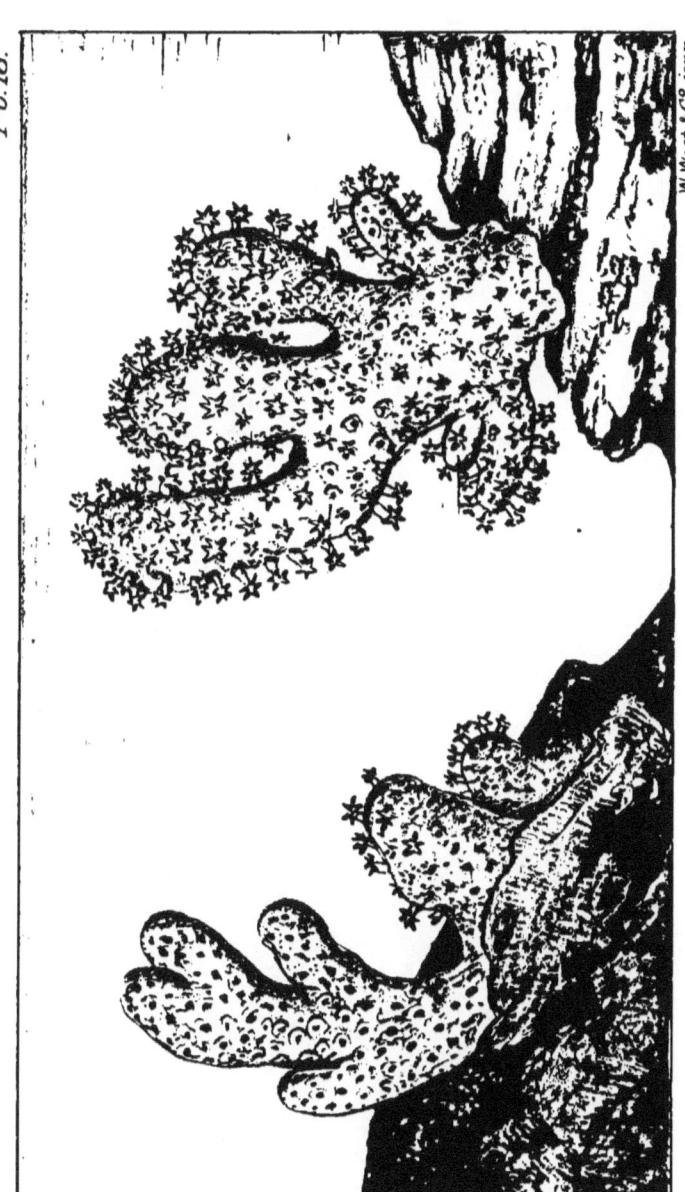

Dead Men's Fingers.

which they are enabled to cling tightly to slippery rocks and stones. The mouth and stomach of the Starfish are situated in the middle of its disc or body. "Small as the mouth of a Starfish appears to be," says a writer on this subject, "small as is its stomach, and feeble as are its muscular powers, it can swallow a bivalve mollusc entire, or, if needful, open it and suck out the contents in some mysterious way, a feat that no man could accomplish without tools. . . . The ancient naturalists were well aware," continues this writer, "that the Starfish possessed the power of eating oysters, but they thought that the creature accomplished its design by watching until an oyster opened its shell, and then poking one of its rays between the shells as a wedge—then, having gained a partial admission, it slowly insinuated itself, and finished by devouring the inhabitant. It appears however, by the reports of careful observers, that the oyster eating is true to the fact, but false as to the mode.

"The Starfish seems to bring its mouth in contact with the edge of the shell, and then from some delicate vessels, never protruded at any other time, to pour into the oyster some drops of a poisonous fluid, which forces the animal to open the shells, and finally kills it. Such is the account as it stands at present."*

There are various species of Starfish, some of them being much more beautiful than others; the kind denominated Five-fingered is one of the commonest, and doubtless has been found on the seashore by most of my readers.

---

* See "Common Objects of the Seashore," pp. 84, 85.

The Webbed Starfish is a very much rarer species: I have never dredged but one specimen, and very beautiful it was—its colour being the most brilliant orange, gradually softening to white under the rays, and deepening into rich scarlet, marked out with black, at the edge or border of the web. The original was twice the size of the drawing, and it was, unfortunately so much bruised and injured, that it died before my sketch was quite finished, and its brilliant tints died with it.

One of the especial peculiarities of Starfish is, that (like crabs, lobsters, and some other Crustacea,) they possess the extraordinary power of breaking off their limbs when handled or otherwise alarmed. Those known by the name of Brittle Stars are particularly prone to dismemberment; the rays of the species are very long and thin, and frequently, when I have taken them in my hand for examination, at least three of the five rays would, in the space of a few moments, be cast off by their alarmed owner.

The tints of these Brittle Stars are extremely brilliant; hundreds, nay, thousands of these creatures has the dredge continually brought up at a time, writhing and entwining themselves together in a gorgeous mass, and yet, owing to their habit of breaking themselves up, I have never been able to preserve one specimen really perfect.

The Sun Starfish is another brilliantly coloured species; its shape is very different from that of the Brittle Star, for instead of having a small body and five long rays, this creature has a comparatively large disc and twelve short arms; its colour is generally pink or scarlet, and, although the Sun-

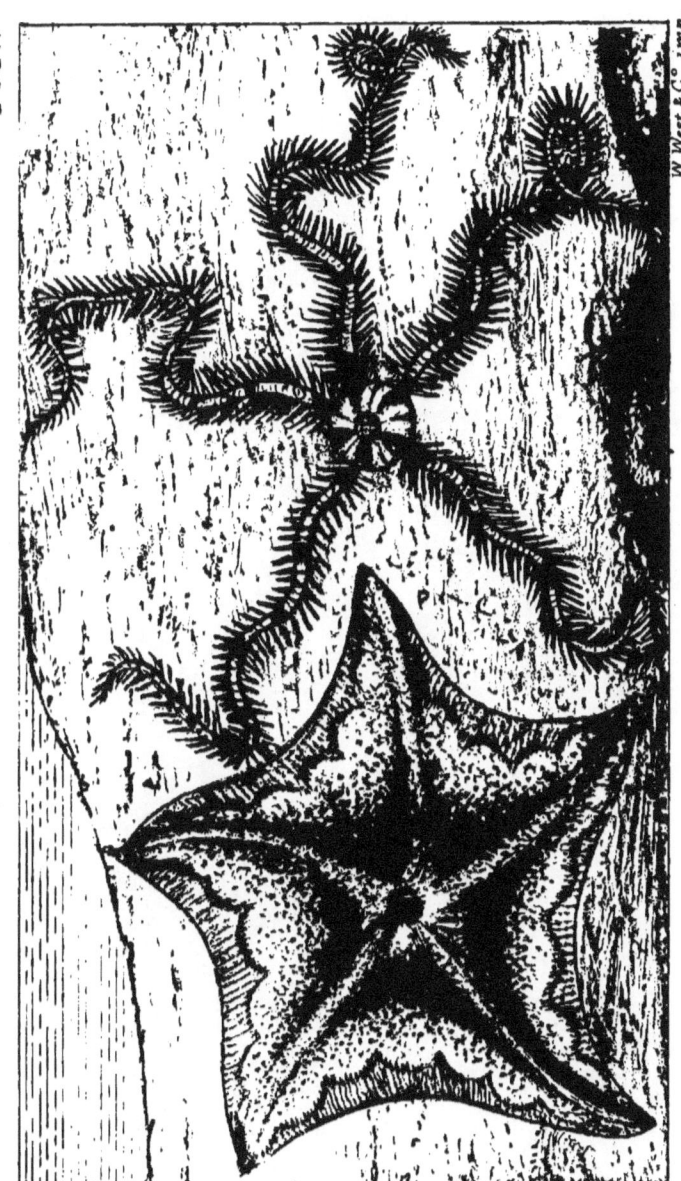

Webbed star-fish.   Brittle star.

star is not a very rare species, it is one of the most beautiful.

The largest British Starfish is the Lingthorn. I have never seen a member of this class, but have read they measure two feet across, and that, like the Brittle Stars, they are extremely liable to break themselves up when angry or frightened, and that, when broken up, "the dismembered fragments continue active long after their dispersion. The feet move about, and attach themselves to any object that comes within their reach, retracting and pushing out with as much vigour as they did when the creature was entire."

I have seen an amusing account of the attempts made by Professor Forbes to capture a specimen of the Lingthorn Star complete. It appears that the first time this gentleman obtained a Star of this description, he was not aware of what he terms "its suicidal powers, and spread it out on a rowing bench, the better to admire its form and colours." And he continues, "on attempting to remove it for preservation, to my horror and disappointment I found only an assemblage of rejected members; my conservative endeavours were all neutralized by its destructive exertions, and it is now badly represented in my cabinet by an armless disc and a discless arm. Next time I went to dredge on the same spot, determined not to be cheated out of a specimen in such a way a second time, I brought with me a bucket of cold fresh water, to which article Starfishes have a great antipathy. As I expected, a Luidia," (or Lingthorn,) "came up in the dredge, a most gorgeous specimen. As it does not generally break up before it is raised above the surface of the sea, cautiously and anxiously I sunk my bucket to a level with the dredge's mouth, and proceeded in the most gentle manner to introduce Luidia to the purer element. Whether the cold air was too much for him, or the sight of the bucket too terrific, I know not; but in a moment he proceeded to dissolve his corporation, and at every mesh of the dredge his fragments were seen escaping. In despair I grasped at the largest, and brought up the extremity of an arm with its terminating eye, the spinous eyelid of which opened and closed with something exceedingly like a wink of derision."

There are many other kinds of Starfish; but as limited space will not allow them to be mentioned, I must pass on to the notice of an inhabitant of the sea, whose outward appearance is, at first sight, very dissimilar to that of the Starfish tribe, but who is, nevertheless, closely related to them. I am speaking of the Echinus, or Sea-urchin, the dead shell of which may frequently be found on the seashore, and is, doubtless, sufficiently familiar to my readers, although a living specimen is seldom washed up by the waves. Like the Starfish, the Sea-urchin is provided with hundreds of sucker feet, capable of being pushed forth and drawn back at pleasure; the position of its mouth is also the same as that of the Star, being in the centre of the under part of its shell, or body, and a terribly large mouth it is, with its five protruding teeth, so particularly hard and sharp.

When alive, the shell of the Sea-urchin is covered with sharp spines, and upon close examination, a number of transparent

Sea-urchin.   Five fingered star-fish.

horns, or feelers, will be seen amongst them; they terminate in a sort of knob, and are continually in motion, being elongated and shortened according to the desire of their owner.

The movements of the Sea-urchin are very slow, in spite of its having such a large number of sucker feet, and being, moreover, probably helped along by some of its numerous spines. It seems to possess a modest and retiring nature, for its chief delight, when in an aquarium, appears to be to cover itself with pebbles, which it picks up with its spines. At first I imagined that the little stones had fallen by mistake, and wishing to do all in my power to render my captive happy, I removed the pebbles with a brush, but the Sea-urchin evidently did not appreciate my would-be kindness, for, in a short space of time, he had again covered himself with pebbles; and so completely was he hidden beneath them, that if he had not crawled up the side of the aquarium with his load, I should have had some difficulty in discovering his whereabouts.

Occasionally the Sea-urchin will add the adornment of a piece of seaweed to that of the small stones, and I was much amused one day to see him moving slowly along with half a mussel shell on his back, (which mussel I had opened, and placed in the aquarium as food for its inmates), and in the half shell sat a tiny crab as comfortably as possible, tearing off bits of the mussel with its claws, and stowing them away in its mouth, as complacently as if a ride on the back of a Sea-urchin were its usual custom at dinner time.

The shell of the Echinus appears, at first

sight, to be moulded in one piece. Such, however, is not the case, several hundreds of pieces, or, as they are properly termed, " pentagonal plates," being joined together, but so neatly, that very careful scrutiny is necessary to observe them at all.

The spines of the Sea-urchin in the illustration are pale green, tipped with purple: the Echinus is also called the Sea-egg, because in some places it is used as an article of food.

And speaking, or rather *writing*, of articles of food, I will now give a few particulars respecting shrimps and prawns; mentioning, in the first place, that the appearance of these delicacies, when alive, is very different from that produced by the influence of boiling water.

Instead of being of a delicate pink as it

appears on our breakfast-table, the prawn, when alive, is of a transparent greenish-grey, streaked with black, white, and orange, intermixed with purple tints: its eyes are extremely bright, and its antennæ remarkably long.

The strong, sharp kind of sword, which projects from the prawn's head, is doubtless used as a defence against its foes, and, judging by the array of teeth on either side of the sword, it must truly be a useful weapon to its owner.

When alarmed, the prawn has the power of darting backwards; this manœuvre is executed by means of its tail, which consists of five fringed pieces, capable of being expanded, or of overlapping each other, at the will of the prawn.

The Crustacean known as the Olive Squat, or the Squat-lobster, is also distin-

Prawn. Shrimp.

guished for being able to shoot backwards through the water by a peculiar movement of its tail.

The Squat is extremely shy and nervous; its long horns and arms are ever on the alert to discover danger, and when alarmed, it throws its arms and legs forwards, and, flapping its tail underneath its body, darts backwards into a hole with wonderful rapidity and precision. The colour of some specimens is dark olive, with light brown or yellow streaks: its claws are rough, and adorned with sundry sharp spikes. Other members of this class are scarlet, and others blue.

The colouring of shrimps is not so varied as that of prawns; their heads are flatter, and devoid of the projecting sword: they are also called Sand-raisers, by reason of their propensity for burying themselves in sand. This they do very quickly by raising up a cloud of sand with their feet, which, falling on them, helps to conceal their whereabouts. Even when not completely covered with sand, it is sometimes rather a difficult matter to find the shrimp, owing to the similitude which exists between its own tints and that of its sandy resting-place.

As before mentioned, the shrimp has not a very happy or long life if placed in a small tank wherein also dwell anemones or crabs, for both races of creatures are always on the watch to catch and demolish the poor little sand-raiser, and, when darting hurriedly away from the outstretched claws of the one enemy, the shrimp is but too often caught in the outspread tentacles of the other. As a rule, the unfortunate victim makes no attempt at escape when in the grasp of a large anemone, such as the Dahlia for instance: it remains as motionless as if it were paralysed, or devoid of life; and yet it is not really incapable of motion, for if removed in good time from the encircling tentacles with the aid of a brush or pencil, it darts away with surprising activity; and if there is any sand provided for it in the aquarium, the rescued shrimp immediately buries itself with remarkable rapidity, considering its late apparent incapability of movement.

Numerous specimens of the creatures commonly called Sea-plumes, or Plumed Slugs, may be found amongst the rocks at low water: the chief characteristic of these Sea-slugs is, that their breathing organs or gills are placed outside their bodies, spread open after the manner of the anemone's tentacles, undefended by shell or any other covering. In consequence of this peculiarity, they are also called Nudibranchiata, or Naked-gilled Molluscs.

The colouring of some Plumed Slugs is extremely delicate : the one represented crawling on a rock, close to the Squat-lobster, was particularly lovely and fragile in appearance, being of a pearly whiteness, whilst the semi-transparent gills or plumes were tinged with pink.

The Sea-slug's mode of progression is the same as that of the Chiton, already described—viz., by means of one foot, attached to the lower surface of its body.

Such, my readers, are a few, and a *very* few, of the countless inhabitants of the sea. Much more could I tell you about them, and many interesting particulars could I

Pl. 20

give respecting that portion of submarine life which especially comes under the denomination of Fish, and of which no mention has been made in the foregoing pages; but I trust that sufficient has already been written to fulfil the object of this little work—viz., to prove that the study and observance of Life Beneath the Waves is fully as interesting, and equally as worthy of attention, as other parts of Natural History.

Such being the case, the construction of the Brighton Aquarium may assuredly be looked upon in the light of a public benefit; for not only will it be a novel and delightful place of resort, and an inestimable boon to naturalists, but the expectation may be safely entertained, that the exhibition of marine animals, in a state of perfect health and beauty, will effectually eradicate apathy and indifference from the mind of the public in general, and establish, instead, feelings of interest and admiration, respecting the beautiful and marvellous creatures which first sprang into existence in obedience to the command of the Almighty Creator of the universe,—

"*LET THE WATERS BRING FORTH ABUNDANTLY.*"

INDEX.

	PAGE
ANEMONES	23
,, Dahlia	24
,, Strawberry	26
Angled-crab	47
Annelids	28
BRIGHTON Aquarium	9
,, ,, site of	11
,, ,, entrance court	12
,, ,, entrance hall	13
,, ,, principal corridor	14
,, ,, second main corridor	17
,, ,, central square	15
,, ,, conservatory	16
,, ,, dining hall	13
,, ,, reservoirs underground	18
,, ,, tanks	15
,, ,, terrace promenades	19
Brittle-star	77
CARPET-SHELLS	70
Chitons	64
,, Marmoreus	65
Crab, edible	38
,, transformations of	39
,, change of shell	42
,, quarrelsome disposition	45
,, power of dismemberment	46
,, Angled	47
,, Helmet	51
,, Hermit	54
,, Nut	48

	PAGE
Crab, Spider	49
„ Swimmer	52
„ Velvet	53
DAHLIA Anemone	24
Dead Men's Fingers	73
Dog-winkle	68
ECHINUS	82
„ shell of	85
Escallops	61
FIVE-FINGERED Star	76
GILLS of Serpulæ and Sabellæ	29
„ Naked-gilled Molluscs	89
HELMET crab	51
Hermit crab	54
„ change of abode	56
„ excessive pugnacity	57
LOBSTER, squat	86
MAIL-shells	64
Molluscs, naked-gilled	89
Mouse, sea	71
„ bristles of	72
NATICAS	66
„ Necklace	67
Nudibranchiata	90
Nut crab	48
OLIVE squat	86
Oyster, starfish's mode of eating	75
PECTEN	58
„ Jacobæus	62

Index.

	PAGE
Pelican's-foot	62
Prawn	86
Purpura	68
SABELLÆ	32
Sea-slugs	89
Sea-mouse	71
Sea-butterfly	59
Sea-urchin	82
Serpulæ	29
Shrimp	88
Squat-lobster	86
Starfish	74
,, Five-fingered	77
,, Brittle	77
,, Lingthorn	79
,, Webbed	76
,, Sun	78
TAPESTRY shells	70
Top-shells, or Trochus	63
Tyre, purple dye of	69
URCHIN, sea	82
VELVET-CRAB	53
Venus, banded	69
WORM, sea, in rock	35
Worm, Scarlet	37
ZOEA	38

THE END.

www.ingramcontent.com/pod-product-compliance
Lightning Source LLC
Chambersburg PA
CBHW021936160426
43195CB00011B/1115